NISTIR 7851

A Testbed for Evaluation of Speed and Separation Monitoring in a Human Robot Collaborative Environment

Sandor Szabo
William Shackleford
Richard Norcross
and
Jeremy Marvel
U.S DEPARTMENT OF COMMERCE
National Institute of Standards and Technology
Intelligent Systems Division
Gaithersburg, MD 20899-8230

March 2012

U.S. DEPARTMENT OF COMMERCE
John Bryson, Secretary
NATIONAL INSTITUTE OF STANDARDS AND TECHNOLOGY
Patrick D. Gallagher, Under Secretary for Standards and Technology and Director

A Testbed for Evaluation of Speed and Separation Monitoring in a Human Robot Collaborative Environment

Sandor Szabo
William Shackleford
Richard Norcross
and
Jeremy Marvel
U.S DEPARTMENT OF COMMERCE
National Institute of Standards and Technology
Intelligent Systems Division
Gaithersburg, MD 20899-8230

Table of Contents

1 Introduction

Collaborative robots are under development throughout the world that will allow humans and robots to operate in close proximity while performing a variety of tasks. A working group[1] within the International Organization for Standardization (ISO) is developing the standards to help ensure these robots operate safely. The international robot safety standard, ISO 10218, consists of two parts: part one which is directed toward the robot manufacturer [1], and part two which is directed toward the robot user [2]. The documents define speed and separation monitoring (SSM) as a form of collaborative robot safety where contact between a moving robot and operator is prevented by limiting robot speed and maintaining an adequate separation distance. The working group is currently developing a technical specification (ISO/TS 15066 Robots and robotic devices – Collaborative Robots) which will provide detailed guidance on collaborative robot operations and safety, including SSM.

The Intelligent System Division (ISD) of the National Institute of Standards and Technology (NIST) is part of the team preparing the SSM portion of technical specification (TS) 15066. This report describes a prototype SSM safety system developed by ISD to support testing and validation of methodologies being developed for the technical specification. The NIST SSM safety system implementation uses laser scanners to measure the position and velocity of humans (or any other moving objects) within the robot workspace and computes the safe separation distance based on the robot's reported position and velocity. The system issues a stop or restart command to the robot controller depending on a minimum separation distance equation proposed in the ISO TS.

The SSM prototype uses a combination of off-the-shelf components, in-house software, and specific procedures including:
- software that uses range data from the scanners to locate and track humans,
- calibration software and procedures to register the laser scanners with each other and with the robot coordinate system,
- robot interface software that outputs the position and velocity of the robot axes to the safety system and accepts commands from the safety system to stop and restart the robot while the robot is running a motion program in automatic mode,
- SSM safety system software that calculates the closing speeds, time-to-contact, separation distance, and control functions for stopping and restarting the robot.
- validation procedures for testing SSM operation and for gathering performance data,
- software for analyzing SSM performance using data from an independent measurement system

This report gives an overview of the SSM system, how it works and how we verified its operation. The system has not undergone safety qualification and is not suitable for a

---

[1] TC 184/SC 2/WG 3 Industrial Safety

1

real-world application. However, several of the results and conclusions are relevant to the preparation of the technical specification and for developing future conformance and validation guidelines.

2   Speed and Separation Monitoring (SSM) Testbed

The Speed and Separation Monitoring testbed consists of a 7-axis under-slung robot, a prototype SSM safety system, a laser scanner-based human tracking system, and a SSM robot interface. Figure 1 shows a top view and a side view of the testbed and the SSM scenario. Each scanner sweeps a laser in a horizontal plane and the tracker uses the range data to isolate moving objects (e.g., humans, guided vehicles). Each object is represented as a circle with identifier, position vector (center of the circle), velocity vector, radius, and time stamp. The robot is represented as four cylinders centered at the tool, forearm, elbow, and base. Each robot cylinder is represented as a circle with position vector (center of the circle), velocity vector, radius, and time-stamp.

The SSM safety system calculates the separation distance between each object circle and robot circles and stops the robot if the distance falls below the minimum separation distance equation currently proposed in the ISO SSM technical specification. The safety system issues a restart to the robot if the distances grow larger than the minimum separation distance. The proposed SSM specification does not take into account the direction of the velocity vector, which means that the robot will remain stopped even while the human is traveling away and will only resume motion once the separation distance exceeds the minimum. Also at this time there is no option to lower the robot's speed in the specification, only to stop the robot. We have included a slow option in our SSM implementation for evaluation in the future.

The robot used in this implementation supports existing robot safety standards where a manual restart (i.e., pressing a button) is required to resume operation of an automatic program after a protective stop. In order to implement the proposed SSM specification with an automatic restart, we implemented a custom interface on the robot's controller. A robot system, when designed in accordance with these developing safety standards will contain the necessary interfaces in its safety system to support a SSM implementation. When describing this implementation we use the term pause and resume interchangeably with protective stop and automatic restart, respectively.

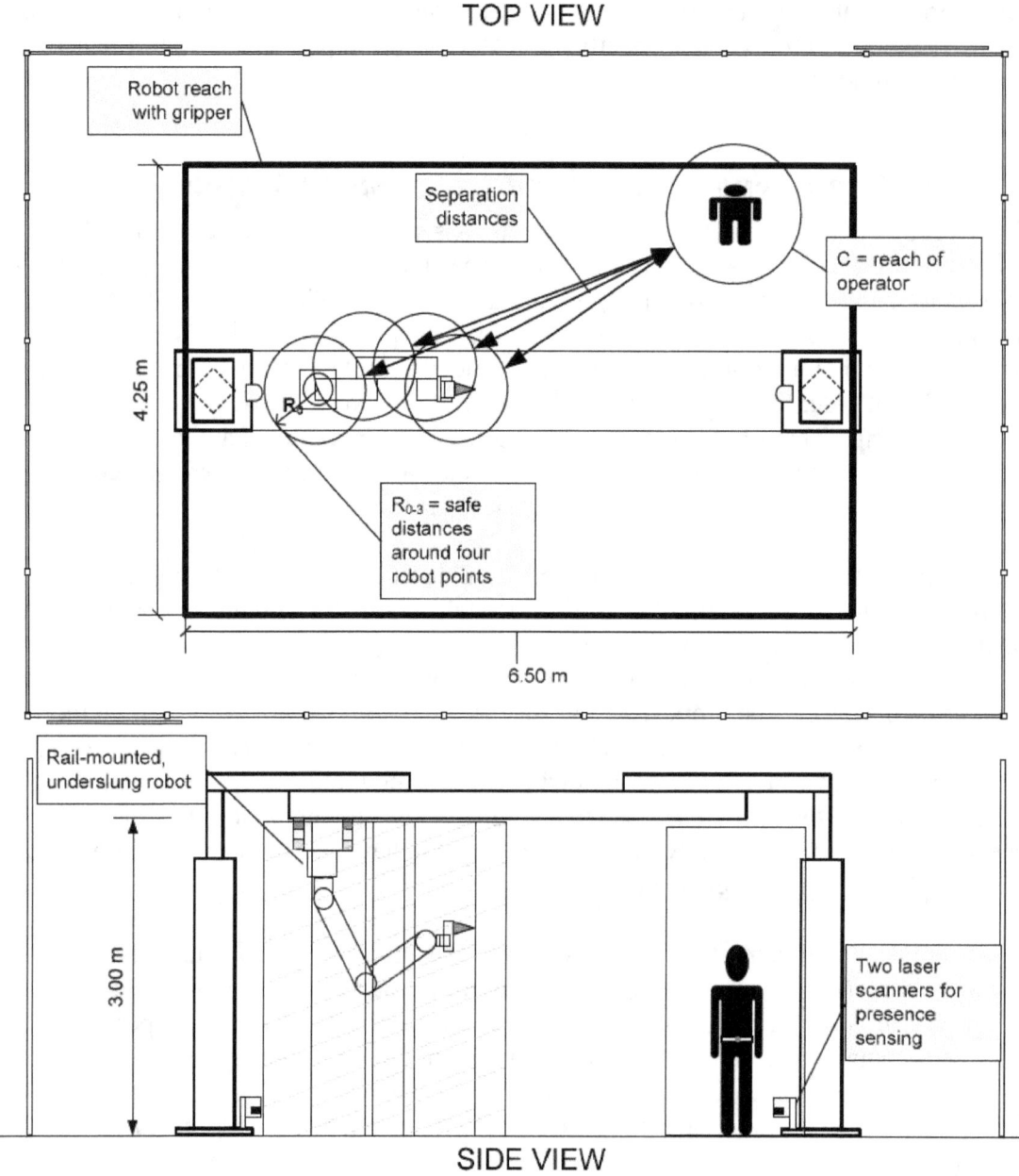

Figure 1. SSM testbed overview

Figure 2 shows an architecture diagram showing the interconnections between the SSM modules and a timing diagram describing the relationship between SSM functions and the robot and human speed. The laser scanners are commercial off-the-shelf devices that produce range and azimuth values from an internal spinning laser. The human tracking module (see Section 3) merges the laser data into a single coordinate system aligned with the robot and detects and tracks moving objects. The SSM controller (see Section 5) computes the separation distance between the robot and the moving objects and issues a pause if the distance falls below the minimum distance. The robot SSM interface (see Section 6) performs two functions. First, it sends the robot tool center point (TCP) position out a socket communication port to the robot tracking module, which creates the four robot circles that represent the robot's position, velocity, and radius. Second, it receives slow, pause, and resume commands from the SSM controller and modifies the robot's automatic program accordingly. Both robot SSM interface functions are written in the robot's host language and reside within the robot controller.

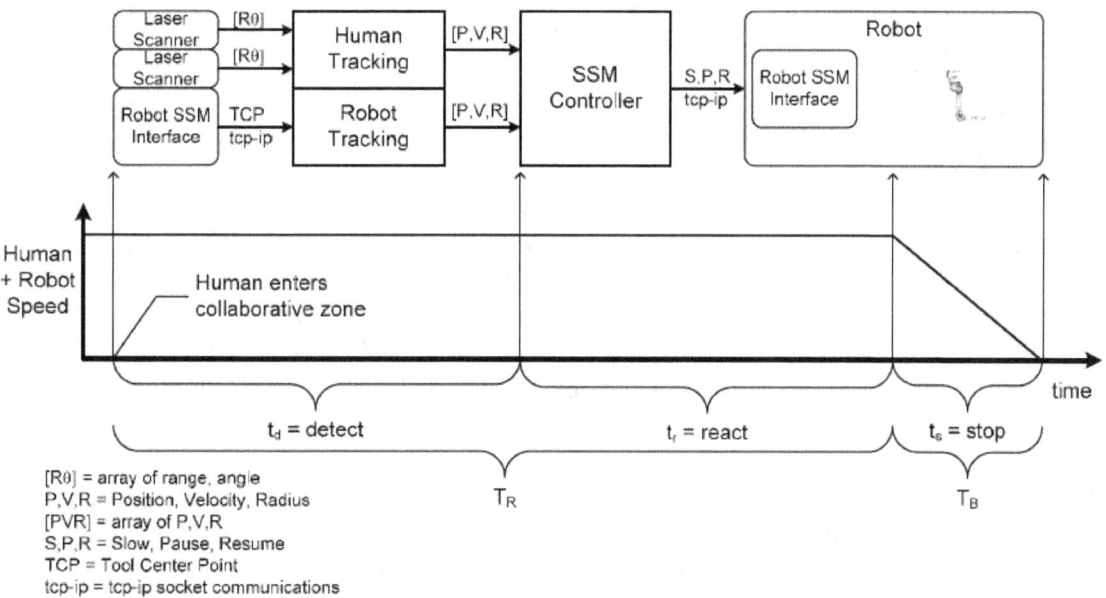

Figure 2. System and timing diagram for SSM prototype

The timing diagram shows the correspondence between the SSM modules and the T_R (reaction time) and T_B (brake time) parameters used in the ISO TS minimum separation distance equation (equation (1) in Section 5). T_R is the time it takes for the SSM system to detect a human and to issue a stop. During this time the SSM assumes the robot and human travel at a constant speed. Note that in our system, T_R also includes the time to get the robot position data. In the ISO TS, T_B is the time it takes for the robot to stop. In our system we subdivide this time into a respond phase (process command) and stop phase (ramp speed down).

3 Human Tracking

The human tracking system is an expanded version of a system we developed for inexpensive ground-truth measurement [3]. The system uses two laser scanners mounted horizontally to the base of each of the columns supporting the rail for the under-slung robot (see Figure 3). The scanners are mounted approximately 0.4 m above the floor facing each other on opposite sides of the robot work volume approximately 5 m apart. The height of the scanners are offset to avoid interference between the lasers. This configuration covers the robot's work area and reduces occlusions due to stationary and moving objects. Also, placing the scanner below the robot's travel eliminates the need to discriminate between a moving object and the moving robot: any motion in the scanner's plane is assumed to be an object to avoid.

4

The tracker combines the range values into a single coordinate system. To accomplish this, the operator must first establish the position and orientation offset between the two sensors. This is done manually by visually aligning on a display the scans produced by sensors. An object is placed in the field of view (FOV) of both sensors. The operator drags the display of the object from one sensor over the display of the same object from the other sensor and rotates the object it until the displays are consistent.

The background is recorded that contains all the static objects in the FOV. Several frames of data are taken and combined to reduce sporadic noise. Objects seen during this background scan, for the current workcell setup, include the legs of a conveyor and the two columns supporting the robot. The tracker detects humans by detecting changes in the static background. Areas where background static objects exist are not processed by the tracker. This eliminates the problem where someone stands still in the robot work volume and eventually is considered background. However, the operator needs to reestablish the background when static objects are moved. Otherwise a human could enter undetected through the previously occupied space. Future work will examine ways to automatically detect changes and automatically update the background.

The human tracking is calibrated to output positions registered in the robot's coordinate system. The procedure uses a 10 cm tube placed in the robots gripper facing down toward the floor. The robot is driven to three widely-spaced positions with the tube low enough to intersect the laser scanner plane. The robot's positions appear on the display along with the tracking system's measurement of the tubes. The operator uses display controls to manually align the robot position and the tube and software automatically calculates the transformation. All subsequent human tracker positions are transformed into the robot's coordinate system enabling the SSM controller to compute the correct separation distances.

During SSM operation, the tracker combines groups of range values into leg components and then groups legs components into human objects. The tracker matches components and objects over time and filters the position of the human using a Kalman filter. The filter assumes that constant velocity will be maintained and can be tuned by setting the expected acceleration variance and measurement variance. The final position and velocity of the human sent to the SSM controller are taken from the estimated state of a Kalman filter.

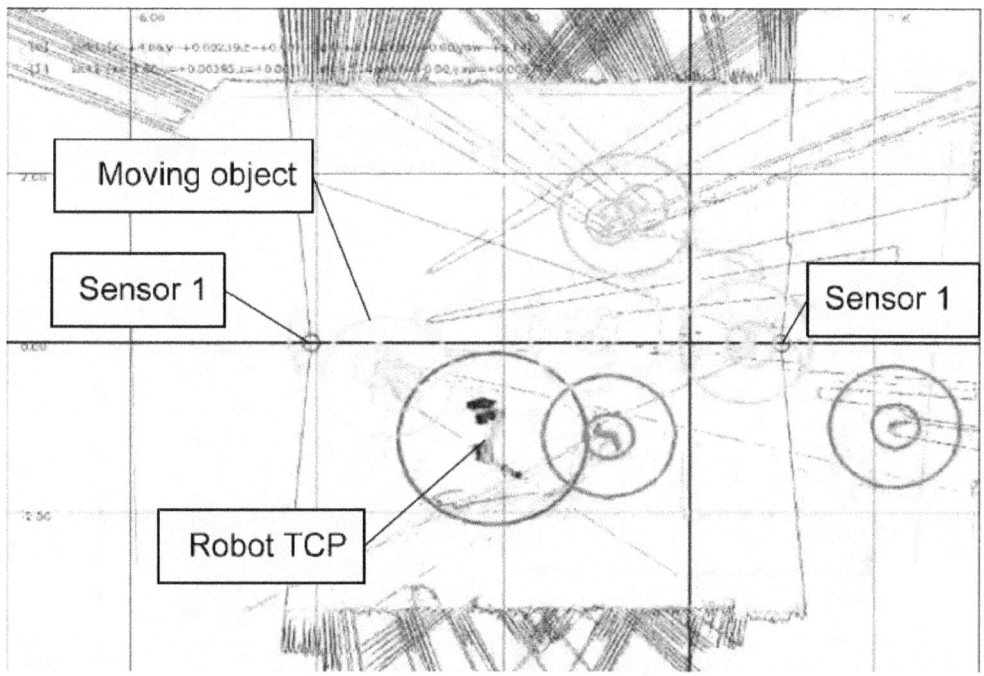

Figure 3. Tracker display showing sensor location, range data, locations of moving objects (colored circles, inner circles are legs) and the location of the robot's tool center point. The radius of a circle is the threshold distance for grouping range values.

One issue being investigated is how to handle occlusions caused by multiple objects or people blocking the laser scanner FOV. These occlusions can mask the approach of other people thereby preventing the SSM from issuing a pause. We extended the tracker to detect occluded regions (see Appendix A), but have not yet designed an interface to pass this information to the SSM controller. The results of the occlusion detection algorithm are shown in Figure 4. The figure shows regions occluded by static objects (yellow) computed from the background range data and regions occluded by dynamic objects (red) computed from the tracking range data.

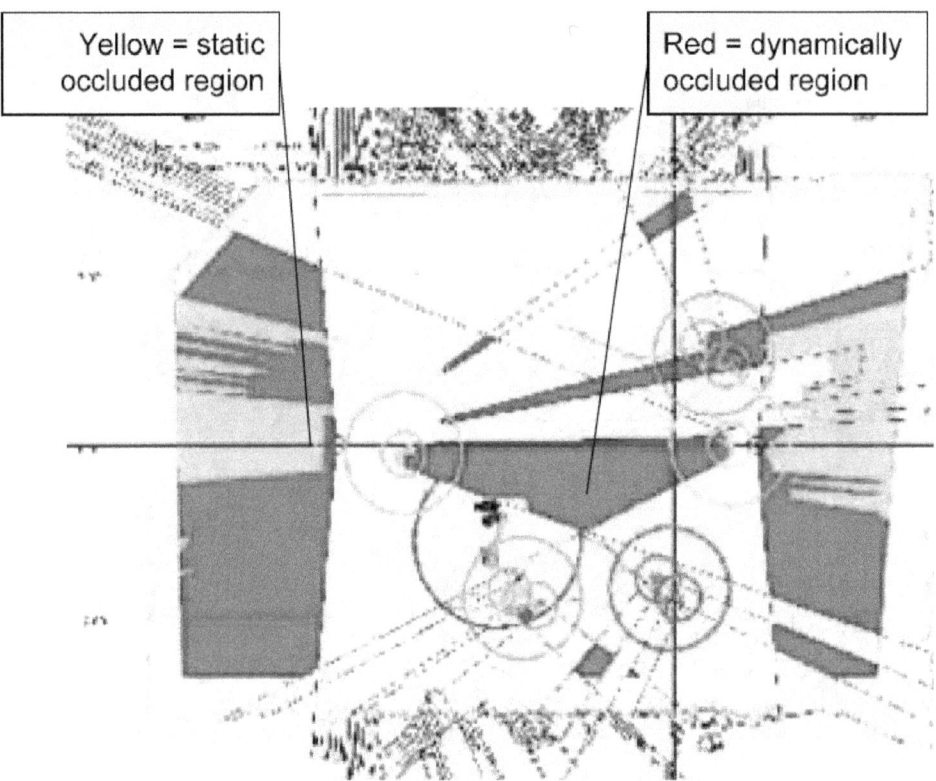

Figure 4. Detecting regions the sensors cannot see due to occlusions by fixed objects and by moving objects.

4   Robot Tracking

A task running on the robot controller determines and transmits the robot position and configuration to a receiving task on an external computer that re-broadcasts the data to other computers and tasks within the testbed. The data consists of the robot's Cartesian position, arm configuration, and joint angles. The Cartesian position is the 6 degree position and orientation of the robot's TCP. The TCP may include offsets associated with the size and dimensions of the tool. The robot configuration identifies the orientation of the shoulder joint (right or left), the configuration of the elbow joint (up or down), the configuration of the wrist, and the wind-up turns of the wrist joints (the final joint range is greater than 360°). Each position message includes a timestamp when the task collected the information from the robot controller.

The position data is sent over a Transmission Control Protocol/Internet Protocol (TCP/IP) socket to a task running on an external computer. The external robot tracking task computes Cartesian and joint velocities based on the position and time information and creates a coarse model of the robot consisting of four X-Y positions. The first position is the location of the robot's TCP. The second is the location of the base along the overhead rail as provided by the robot's joint position. The third and fourth locations are close to the robot's forearm and elbow respectively. The SSM system uses the positions and additional radius parameters to represent the robot and to calculate the separation distances between parts of the robot and objects in the collaborative workspace.

5 SSM Controller

The SSM controller's primary function is to monitor speed and separation distance between the robot and the operator and to issue a stop to the robot prior to contact. A presence sensing device is often used in traditional robot applications to detect a person in time to stop the robot prior to contact. Robot safety engineers use ISO 13855 "Safety of machinery — Positioning of protective equipment with respect to the approach speeds of parts of the human body" to determine where to position the presence sensing device. In these applications, the sensing device is situated outside the robot's work volume and the robot will stop prior to the human entering inside the work volume. However, in a collaborative application the human can enter areas inside the work volume (termed the collaborative space) while the robot is in motion. In order to stop the robot before contact, the equation in ISO 13855 is modified to account for the robot's speed and dimension. Equation (1) shows the collaborative form of the minimum separation distance equation.

$$S = K_H(T_R + T_B) + K_R(T_R) + B + C \tag{1}$$

Where:

K_H	=	Speed of human
K_R	=	Speed of robot
T_R	=	Reaction time to detect human and issue a stop – a parameter measured during timing test (see Appendix B).
T_B	=	Brake time – see below
B	=	Brake distance – see below
C	=	$C_H + C_R$, the region surrounding the human and robot respectively. For the testbed, this region includes the uncertainty in position and dimension of each.

For the SSM testbed, the brake distance is:

$B = \frac{K_R^2}{2A}$

where A is a worst-case deceleration level measured during stopping tests (see Appendix B), and the brake time is:

$T_B = K_R / A$.

The SSM calculates the distance, D, between each object detected by the tracker and each point on the robot. The SSM issues a stop if the distance minus the minimum separation distance, S, goes negative (see Figure 5).

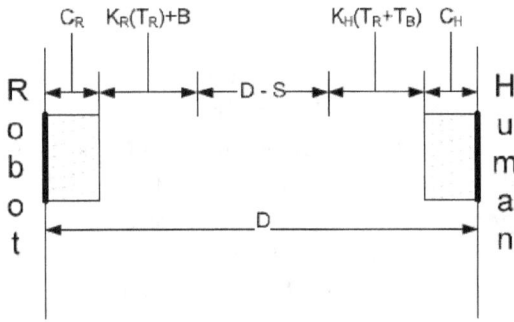

Figure 5. The SSM issues a stop when $(D - S) < 0$.

The SSM technical specification allows the use of static or dynamic human and robot speeds: static speeds use fixed maximum defaults and dynamic speeds use actual measured speeds. The SSM controller (see Figure 6) displays circles around points on the robot and around moving objects. The robot inner circle radius is C_R and the outer circle radius is $K_R T_R + B + C_R$. The object inner circle radius is C_H and the outer circle radius is $K_H (T_R + T_B) + C_H$. In dynamic mode, the outer circles grow based on the current speeds of the robot and object, and in the static mode the circles remain fixed based upon maximum fixed speeds, 2.0 m/s for a human and the programmed speed for the robot. The SSM issues a stop whenever a robot's outer circle comes into contact with an object's outer circle and the robot should stop before the inner circles come into contact.

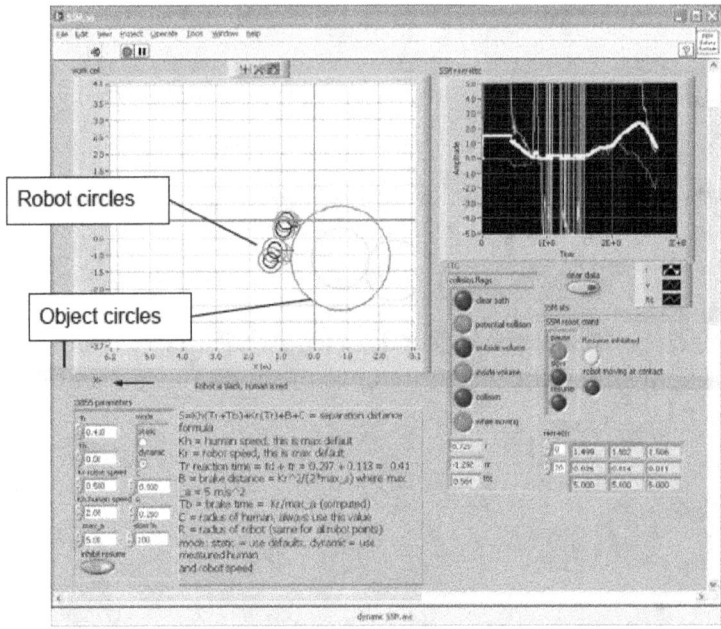

Figure 6. SSM display shows circles giving position of the robot and moving objects. The outer circles in red define boundaries around the robot and the object. A stop is issued if the circles touch.

9

6 Robot SSM Interface

Commercial robot manufacturers do not currently incorporate an SSM interface in their controllers. For our testbed, we implemented an SSM interface via two robot background tasks run on the robot controller: an SSM control task and an SSM communications task. The tasks may be initiated at robot power-up or manually. The SSM communications task accepts commands from an external computer via a TCP/IP socket and sets an internal command variable that indicates the robot should resume, slow, or stop. The SSM control task monitors the command variable along with three pairs of digital input lines (six lines total) that provide an alternate method for external devices to issue these commands. In addition, the SSM control task monitors a binary mask that allows the system to ignore individual signals.

The SSM communications task establishes a socket link at initialization and then waits for external commands. The SSM control task stops the robot by first polling the tasks searching for the motion task and then pausing the task if it was not previously paused. The SSM control task indicates a stop state by illuminating a red indicator lamp on the robot arm. If none of the stop signals are asserted or a resumed command is received, the SSM control task releases the paused motion task and the robot resumes motion.

The SSM control task is also used to set the robot controller's general speed override variable to implement the SSM slow command. The general override value reduces the speed based on a percentage of a commanded speed. For example, if the robot program commands 1000 mm/s and the general override is 50%, the robot will move at 500 mm/s. The SSM control task limits the robot speed to a maximum value as opposed to reducing all of the robot's programmed speeds to avoid excessively slow speeds. The SSM task monitors the robot's commanded speed via a controller system variable. When the commanded speed exceeds the speed limit, the SSM task applies the appropriate general override to bring the robot to the limit speed (override = speed limit/commanded speed). The SSM task illuminates a yellow indicator lamp on the robot arm to indicate a slow state. When none of the stop or the slow signals are asserted, the SSM task sets the general override to 100% and illuminates the green light on the robot arm. We measured the interface reaction time using the procedure described in Appendix B.

7 Validation Testing

The purpose of validation testing is to determine if the robot comes to a complete stop prior to the separation distance falling below $C_R + C_H$. For safety purposes, we use a mannequin mounted to a rolling platform. The mannequin's legs and spacing replicates the anatomy of a human, but does not provide the walking motion. Due to the dynamic nature of the measurement, i.e., the human may be moving at the time the robot stops, we use an external independent measurement system described in Section 8 to track the robot and mannequin position simultaneously. We mount a target in the robot's gripper and a target at the center of the mannequin (see Figure 7). The robot runs back and forth between two positions approximately 2 meters apart at various speeds, from 0.5 m/s to 2 m/s. We push the mannequin back and forth in the direction of the robot while the robot is moving. We record the position of the targets during the test and analyze the performance off-line. Section 9 gives examples of validation testing results.

Figure 7. Validation test set-up. The robot and mannequin have separate targets that are tracked using an independent measurement system.

8   Independent Measurement System (IMS)

The IMS used for the validation tests is an inexpensive, commercial–off-the-shelf system that uses infrared cameras to track retro-reflective spheres 1 cm in diameter (see Figure 7). A target consists of multiple spheres arranged in a unique pattern. The system uses triangulation to measure position and the unique pattern of the spheres to recognize and track the targets.

Our system consists of six cameras. The entire work-volume of the robot cannot be covered adequately by the cameras in any single position. However the cameras can be moved and recalibrated so as to provide adequate coverage over the section of the robot work volume that will be of interest for a given test. Good coverage usually requires that the target be facing at least three of the six cameras. The cameras must also be within range of the target. The range depends on the exposure and threshold settings for the camera. The camera can be made more sensitive if there is nothing in the environment reflective to infrared or very close to the camera. There is also some capability to mask out isolated static reflective objects without requiring the camera to be made less sensitive. The most common consequence of inadequate coverage is that the target will tend to drop out and no position information will be available for that target for some period of time. There is also some risk of one target being mistaken for another, which might cause the robot position to be reported as the mannequin position or vice-versa. This should be corrected later in the analysis. Tests we conducted showed the system had a maximum error of 5.3 cm (see Appendix C).

9 SSM Test Results

In August and September 2011 we evaluated the performance of the SSM safety system on the NIST robot. For these tests the minimum separation distance equation (1) (where a stop is issued) parameters were:

K_R = 0.5 m/s

T_R = 0.41 s.

A = 5.0 m/s^2

T_B = K_R/A

B = $K_R^2/2A$

C_H = 0.9 m

C_R = 0.25 m

In the static trials, the velocity of the mannequin is the worst-case value specified in ISO 13855: $K_H = 2.0$m/s. The minimum separation distance is:

$$S = 2.0\text{m/s}\left(0.41\text{s} + \frac{0.5\text{m/s}}{5.0\text{m/s}^2}\right) + 0.5\text{m/s}(0.41\text{s}) + \frac{0.5^2\text{m/s}}{2 \cdot 5.0\text{m/s}^2} + 0.9\text{m} = 2.15\text{m}$$

Figure 8 shows the result of a trial consisting of four approach events (an approach event is where the mannequin is pushed toward the robot). The data in the figure comes from the independent measurement system. The "TCP Velocity" (see legend) is the speed of the target mounted in the robot's gripper. The velocity is signed to indicate the robot changed direction (recall the motion is back and forth approximately 2 m along a straight line). The "Distance" is the distance between the target centered in the robot's gripper and the target centered on the mannequin. The "Pause Threshold" is the minimum separation distance. The SSM should issue a pause if the distance falls below the pause threshold. The "Contact Threshold" is $C_R + C_H$ (1.15 m). The SSM fails if the distance falls below the contact threshold and the robot is moving (TCP velocity greater than zero).

12

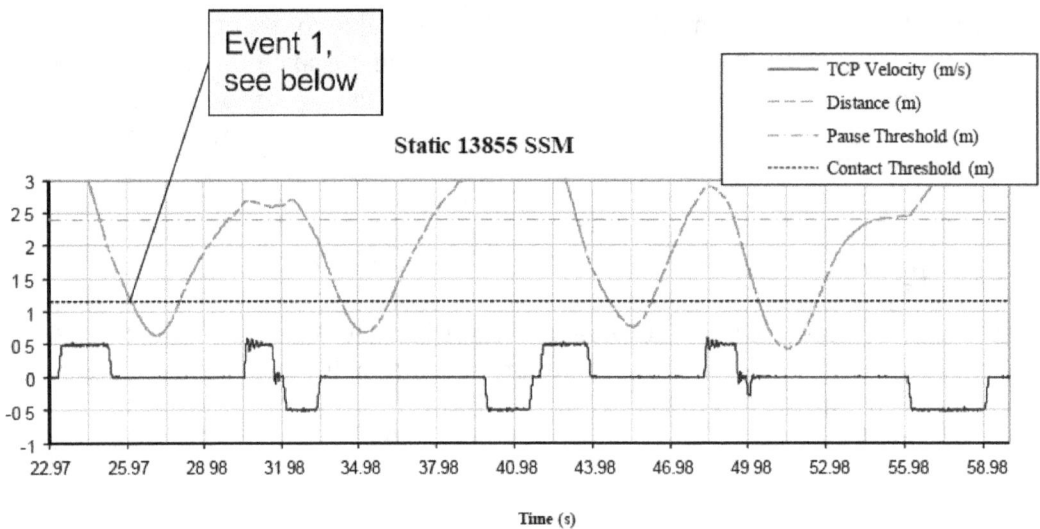

Figure 8. Plot of the robot's TCP velocity (solid blue line) juxtaposed with the distance (dotted red line) between the TCP with the mannequin's center of mass for the entire Static 13855 SSM trial.

A close-up view of one of these events is shown in Figure 9. We do not know the time that SSM issued a stop; however, we expect a stop is issued once the separation distance drops below the pause threshold (see Issue Stop call-out). The remainder of the TCP velocity profile shows a constant speed for approximately 0.4 s (T_R) and then a ramp down in speed with the robot stopping prior to the distance reaching the contact threshold. An examination of all the events showed the SSM performed as expected.

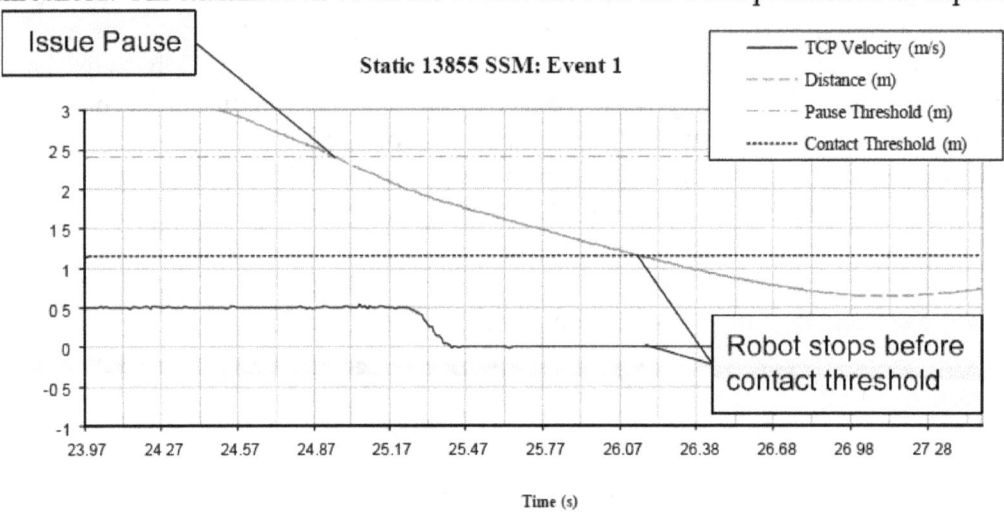

Figure 9. Close-up of Event 1.

In the dynamic trials, the SSM uses the measured human and the measured robot speeds with the remaining parameters the same as in the static case:

$$S = K_H \left(0.41s + \frac{0.5\text{m/s}}{5.0\text{m/s}^2} \right) + K_R(0.41s) + \frac{0.5^2\text{m/s}}{2 \cdot 5.0\text{m/s}^2} + 0.9\text{m}$$

We tested three robot speeds: (0.5, 1.0, and 2.0) m/s. Figure 10 shows the result of four approach events during a test where $K_R = 0.5$ m/s. The pause threshold varies as a function of K_H and K_R.

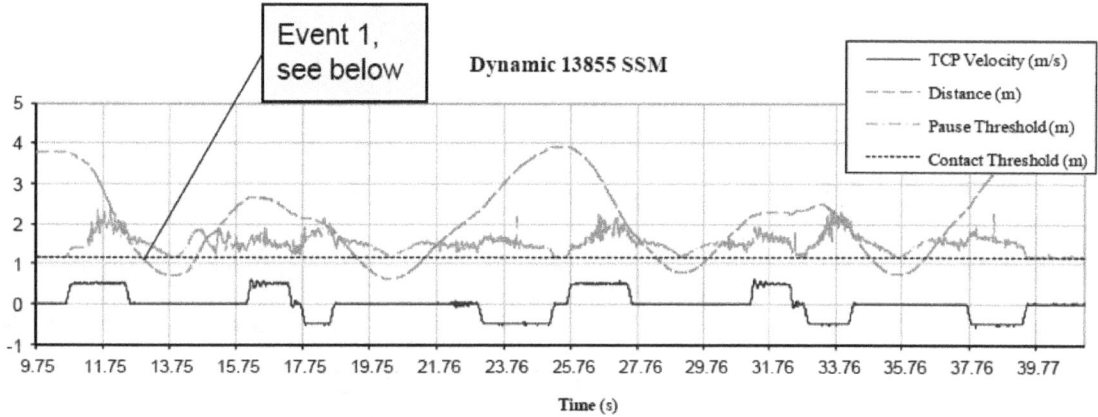

Figure 10. Plot of the robot's TCP velocity (solid blue line) juxtaposed with the distance (dotted red line) between the TCP with the mannequin's center of mass for the entire Dynamic 13855 SSM trial.

Figure 11 shows a close up of event 1. Once again, the SSM performs as expected.

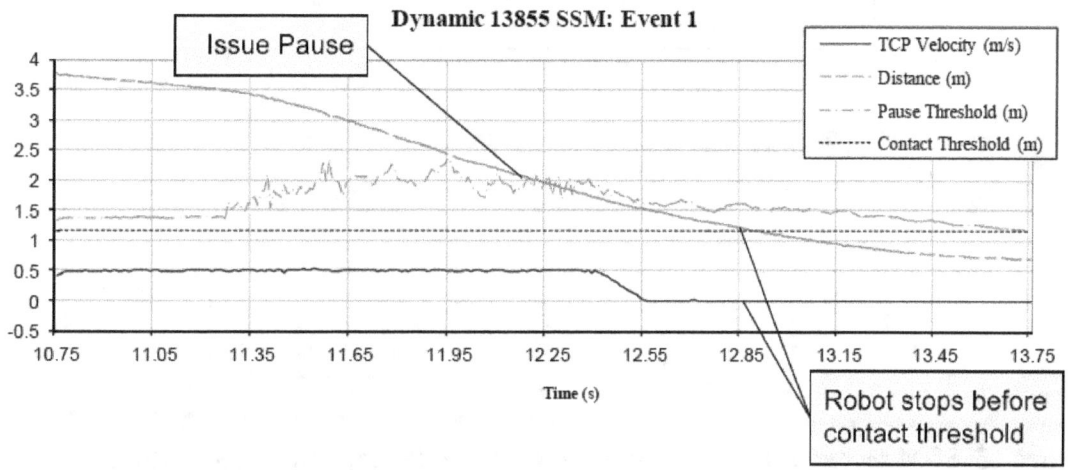

Figure 11. Close-up plot of the robot's TCP velocity and the separation distance for the third of four encounter events during the Dynamic 13855 SSM trial.

We also ran validation tests of the dynamic SSM with the robot traveling at 1.0 m/s (see Figure 12) and at 2.0 m/s (Figure 13). The SSM stopped prior to contact at 1.0 m/s but was not fully stopped prior to contact at 2.0 m/s. The distance overshoot could be attributed to uncertainties in the mannequin's position.

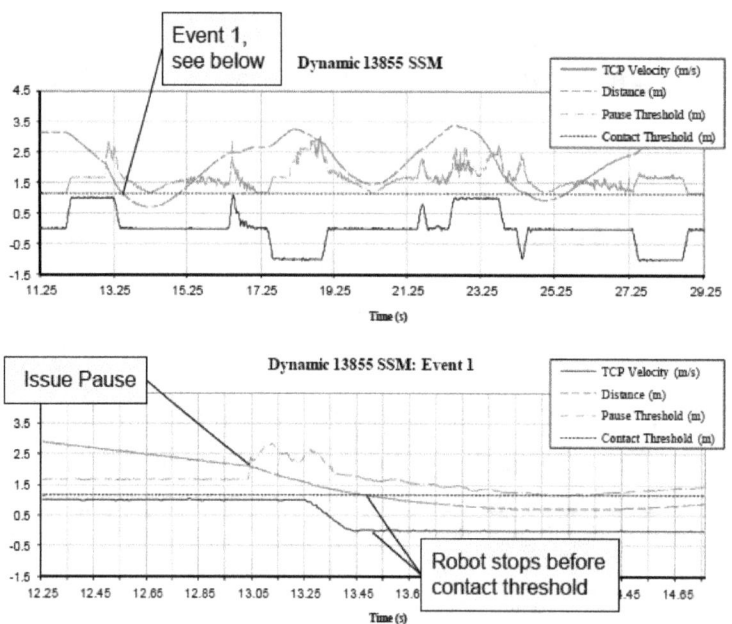

Figure 12. SSM test at 1.0 m/s – Robot stops well before contact

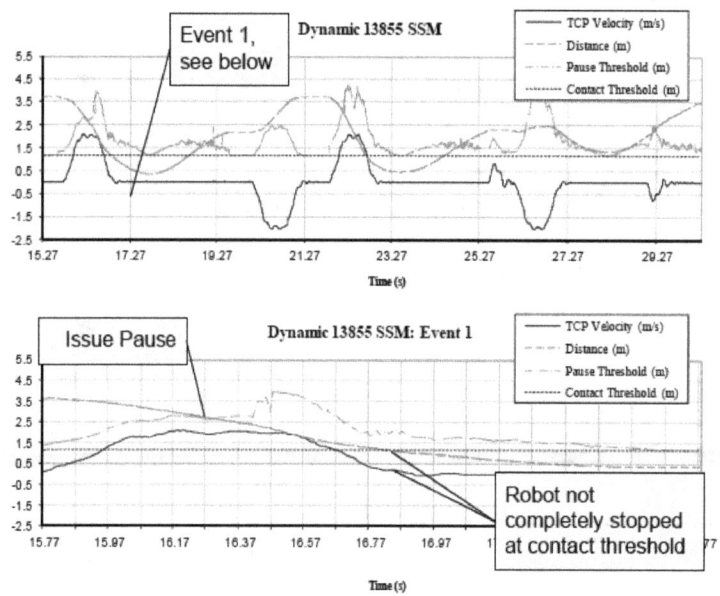

Figure 13. SSM test at 2.0 m/s – Robot not completely stopped when contact threshold reached

10 Future Work

Several variations of SSM are under development that takes into account the direction of travel and the closing speed. The availability of position and velocity vectors enables straightforward calculation of closing speed and time-to-contact (see Appendix D). Using direction and closing speeds is useful when a human hand guides a robot, or when a human and robot move a part together.

The latest robot safety standard does describe an option to lower the robot speed during collaborative operation, it only prescribes stopping. We experimented with a speed reduction option if the operator enters within a certain distance from the robot. Besides reducing the required separation distance (recall the equation uses robot speed), the speed reduction also give the operator feedback that the tracking system is working.

Other forms of preventing contact are being studied. Spencer *et al* [4], for example, utilized first- and second-order instantaneous approximations for time-to-collision of moving objects to drive collision detection, end-effector velocity scaling, and coordinated null-space optimization across multiple affected robots in a shared space. Kulić and Croft [5] utilized a danger index that was a function of inertia and separation distance in their multi-tiered safety system that incorporated long, medium, and short term safety goals. While still other implementations of SSM effectiveness have incorporated factors such as impact force [6]-[8], velocity and robot configuration [9], the most common metric for assessing SSM effectiveness is the Euclidean distance separating bounding volume primitives representing the robot and any detected obstacles in the work environment.

Planned safety metrics research at NIST draws inspiration from the aforementioned historical bases of separation effectiveness, but will also take into account the tradeoffs associated with improving productivity in a manufacturing environment. For example, a robot standing still is very safe yet very unproductive. Metrics to evaluate the tradeoffs between degrees of safety and productivity may prove valuable to the acceptance of human robot collaboration in industrial applications.

11 References

[1] ISO 10218-1 Robots and robotic devices – Safety requirements – Part 1: Industrial robots, July 1,2011, www.iso.org

[2] ISO 10218-2 Robots and robotic devices – Safety requirements – Part 2: Industrial robot systems and integration, July 1,2011, www.iso.org

[3] W.P. Shackleford, T.H. Hong, T. Chang, "Inexpensive Ground Truth and Performance Evaluation for Human Tracking using multiple Laser Measurement Sensors." *Proceedings of the 2010 Performance Metrics for Intelligent Systems (PerMIS) Workshop.* 2010. http://www.nist.gov/manuscript-publication-search.cfm?pub_id=906630

[4] Spencer, *et al.* "Collision Avoidance Techniques for Tele-Operated and Autonomous Manipulators in Overlapping Workspaces." *Proceedings of the IEEE International Conference on Robotics and Automation.* 2008. Pp. 2910-2915.

[5] D. Kulić & E. Croft. "Pre-Collision Safety Strategies for Human-Robot Interaction." *Autonomous Robots.* 2007. Vol. 22, No. 2. Pp. 249-164.

[6] M. Nokata, K. Ikuta & H. Ishii. "Safety-Optimizing Method of Human-Care Robot Design and Control." *Proceedings of the IEEE International Conference on Robotics and Automation.* 2002. Pp. 1991-1996.

[7] K. Ikuta & M. Nokata. "Safety Evaluation Methods of Design and Control for Human-Care Robots." *International Journal of Robotics Research.* 2003. Vol. 22, No. 5. Pp. 281-297.

[8] S. Haddadin, A. Albu-Schäffer & G. Hirzinger. "Requirements for Safe Robots: Measurement, Analysis and New Insights." *International Journal of Robotics Research.* 2009. Vol. 28., No. 11-12. Pp. 1507-1527.

[9] Lacevic & P. Rocco. "Kinetostatic Danger Field – A Novel Safety Assessment for Human-Robot Interaction." *Proceedings of the IEEE/RSJ International Conference on Intelligent Robots and Systems.* 2010. Pp. 2169-2174.

[10] Aiguo Li, Lin Wang, and Defeng Wu. "Simultaneous robot-world and hand-eye calibration using dual-quaternions and Kronecker product." International Journal of the Physical Sciences. 2010. Vol. 5(10). Pp. 1530-1536.

Appendix A <u>Occlusion Detection Algorithm</u>

To find the occluded areas the tracker creates a bidirectional graph network. Each node in the graph contains the location point where the laser was reflected and a node number obtained by incrementing a global count as each node is added. The node is connected to the sensor location for the first and last element in each sensor's range scan, and otherwise to the next and previous node. The size of the graph is reduced by combining consecutive nodes of approximately equal range from the sensor. The size of the graph is also reduced by combining all consecutive points outside a manually chosen protected area polygon. The system creates a graph for each sensor. The graphs are combined by searching for intersecting nodes. At each intersection the connections between the original nodes are broken and all involved points are connected to the new node at the intersection. The combined graph is searched to find all polygons. Too find a polygon, begin at any node, and then traverse to any node it is connected to. After the first move always choose the next connected node by choosing the connected node with the smallest possible angle to the previous node. Repeat until you return to the starting node. If you go to every node and apply this to every connection you will have many polygons stored redundantly. For example, the polygon found starting at node 1 of 1,3,6,7,8 would also be found by starting at 6 as 6,7,8,1,3. To eliminate these redundancies each polygon is normalized by starting the polygon at the minimum node number. The polygons can therefore be compared to eliminate the redundant ones. The outer-most polygon (an artifact of the sensing system) will also be found this way and is eliminated by testing any point not on the edge of the polygon to determine if it is inside the polygon. Each polygon in the list is labeled occluded or not by testing one internal point to determine if the polygon is visible to at least one of the line scanners. The internal point is computed by averaging three consecutive points in the polygon with an internal angle less than 180°. The point is tested by comparing its distance to each sensor with the range reported by that sensor at the appropriate angle.

Appendix B <u>Robot Reaction Time</u>

A Speed and Separation Monitoring (SSM) system stops the robot prior to an impact with a human. Sensors determine the robot's velocity (V_R) and the human's velocity (V_H). Then, based on the sensor detection time (t_d), the robot reaction time (t_r), and the robot stopping time (t_s), the SSM determines the stopping distance (hashed area below) and issues a stop signal.

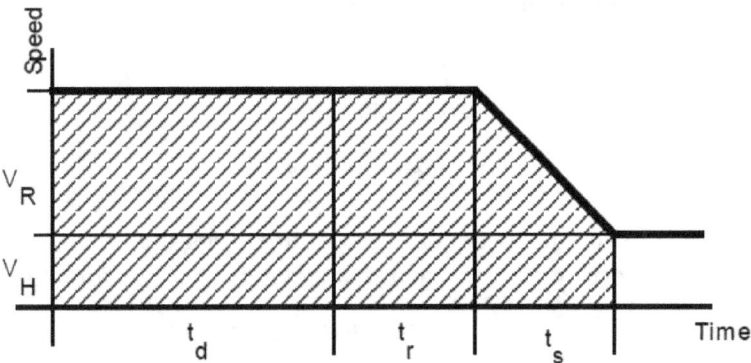

Figure 14. SSM Stopping Diagram

The detection and reaction times vary based on the peculiarities of the sensor and robot. None of the times can be accurately measured by direct observation without access to the inner workings of the robot controller. However, the robot's stopped position is easily determined and can indicate the robot's acceleration and the detection and reaction times.

Robot Parameters

The stopping equation is:

$$d = (V_H + V_R)(t_d + t_r + t_s) - \tfrac{1}{2}at_s^2$$

Where a is the acceleration of the robot which is assumed to be constant. The stopping time (t_s) is the time required to go from the velocity to zero and is a function of the acceleration and the robot's velocity. We determine the robot parameters with measurements of a stationary human ($V_H=0$). Thus the equation is rewritten:

$$d = V_R t_d + V_R t_r + \frac{v_R^2}{2a}$$

To determine the values of the individual parameters, we separate the sensor and robot components. With a fast detector, the detection time (t_d) becomes effectively zero. For example, a beam reflector sensor is not a suitable SSM sensor but has a 0.3 ms response, can generate a SSM-like signal, and effectively isolates the robot components. With zero detection time, the stopping equation becomes:

18

$$d = V_R t_r + \frac{V_R^2}{2a}$$

While the exact robot position at the event signal is difficult to ascertain, the stop positions from different speeds are readily measured. The difference between the stop positions is equivalent to the difference between the stopping distances. The stopping distance difference between the robot at full speed (100%) and at some lesser speed ($i\%$) is:

$$\Delta d = d_{100} - d_i = V_{100} t_r + \frac{V_{100}^2}{2a} - V_i t_r - \frac{V_i^2}{2a}$$

For this analysis, we assume the response time and the acceleration are constants. We also define the velocities as a percentage of the maximum velocity.

$$V_i = i V_{100}$$

The stopping distance difference equation becomes:

$$\Delta d = \left(V_{100} t_r + \frac{V_{100}^2}{2a} \right) - (V_{100} t_r) i - \left(\frac{V_{100}^2}{2a} \right) i^2$$

The stopping distance is the difference between the position along the trajectory when the stop event occurs (P_e) and the position along the trajectory where the robot comes to rest (P_s).

$$d = P_e - P_s$$

The change distance differential is:

$$\Delta d = d_{100} - d_i = P_{e,100} - P_{s,100} - P_{e,i} + P_{s,i}$$

For these tests, the event positions are constant. Therefore:

$$P_{s,i} = \left(V_{100} t_r + \frac{V_{100}^2}{2a} + P_{s,100} \right) - (V_{100} t_r) i - \left(\frac{V_{100}^2}{2a} \right) i^2$$

Thus the relationship between the stopping position and the speed ratio is a quadratic function. The equation shows the quadratic coefficients are functions of the robot parameters.

$$c_0 = V_{100} t_r + V_{100}^2/2a + P_{s,100}$$
$$c_1 = V_{100} t_r$$
$$c_2 = V_{100}^2/2a$$

Regression analysis determines the quadratic coefficients of the underlying data. We measure the maximum velocity separately and use c_1 and c_2 to determine the reaction time and the acceleration.

$$t_r = c_1/V_{100}$$

$$a = V_{100}^2/2c_2$$

Sensor Parameters

Once the robot's parameters are determined, we repeat the measurements with the SSM sensor system and compare the stopping distances as a function of the robot velocity. Since we measure the robot positions, and not the robot to human distance, the human velocity (V_H) is irrelevant. As shown in the figure below, the distance the robot travels during sensor detection is the total stopping distance minus the robot stopping distance.

19

The sensor detection time is the difference between the total stopping distance and the robot stopping distance divided by the robot's velocity.

$$t_d = (d_T - d_R)/V_R$$

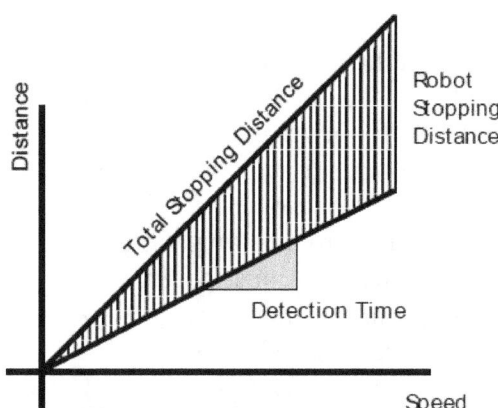

Figure 15. Detection Time Concept

As with the robot parameters, we measure the final stop position more easily and accurately than the distance. We reorder the equation to:

$$d_T - d_R = t_d V_R$$
$$(P_{T,s} - P_{T,e}) - (P_{R,s} - P_{R,e}) = t_d V_R$$
$$(P_{T,s} - P_{R,s}) - (P_{T,e} - P_{R,e}) = t_d V_R$$
$$P_{T,s} - P_{R,s} = t_d V_R + (P_{T,e} - P_{R,e})$$

The event positions are similar for the robot test and the robot and sensor test. Small variations between the event positions affect the intercept but not the slope of linear relationship. The detection time is the slope of the linear regression between the velocities and the difference between the stopping positions.

Uncertainty

We use the standard errors of the coefficients of reduction to determine the uncertainties of the sensor and robot parameters. Computation packages include libraries that compute both the regression coefficients and the confidence interval of the reduction coefficients. The confidence interval represents the extreme value to be expected from results within a percentage of all events. Since the SSM is a safety system, we use the relatively high confidence level of 0.999.

The parameters are functions of the reduction coefficients. To find the confidence interval of the parameter, we add the confidence interval to the coefficient, apply the parameter function, and subtract the parameter computed with the coefficient alone.

$$CI = f(c_i \pm CI_{c_i}) - f(c_i)$$

We are interested in only one side of the Confidence Interval. We desire the largest reaction and detection times and the smallest acceleration. Therefore the coefficient Confidence Interval is added for the reaction and detection times (c_1), and subtracted for the acceleration (c_2).

20

Method

A simple test sequence generates the data required to compute the robot and sensor parameters.

1. Program the robot to move along a linear path.
2. Mount a reflector on the robot.
3. Mount the sensor on a stand in the robot volume.
4. Connect the sensor output to the robot's inputs used for SSM.
5. Build a robot trajectory that passes the sensor.
6. Run the robot trajectory and record the robot stop position.
7. Repeat step 6 for multiple repetitions at various speeds.
8. Determine the robot's positions along the trajectory.
9. Compute the robot parameters as discussed above.
10. Connect the SSM sensor.
11. Set up the intruder at the trigger distance from the original sensor's stand.
12. Repeat steps 6 and 7.
13. Compute the sensor parameters as discussed above.

Results

The results below were collected at the NIST robot testbed. The test limited robot motion to the linear rail to which the robot is mounted. The SSM algorithm executed as a background task on the robot's controller. Lasers scanners measured the position and velocity of objects. The sensor results were relayed to the robot controller via a sequence of TCP/IP links. We made no effort to optimize the data flow between the sensor and the robot motion. Therefore, the resulting parameters are greater than would be acceptable in actual applications.

The data in Table 1 gives the normalized stop positions along the robot trajectory. At higher speeds the robot approached the end of its trajectory with the sensor and data could not be collected. Table 2 shows the computed values and the SSM parameters. The slow reaction times reflect the priority of the background tasks on the controller.

Table 1 Robot Parameter Data

Speed (mm/s)	Speed (%)	Beam Break Stop Position (mm)	Sensor Stop Position (mm)
1	0	0.0	0.0
100	5	11.7	11.3
200	10	25.8	35.3
300	15	42.2	54.1
400	20	59.1	79.3
500	25	80.6	84.5
750	38	141.3	139.8
1000	50	215.1	193.2
1250	63	300.1	308.9
1500	75	400.1	381.8
1750	88	500.8	
2000	100	628.1	

Table 2 Parameter Results

Parameter	Value	0.999 Confidence Interval
c_0	628.2	
c_1	2.26	
c_2	0.04	
V_{100}	2000 mm/s	
t_d	0.251 s	+0.018 s
t_r	0.113 s	+0.019 s
a	5000 mm/s^2	-450 mm/s^2

Appendix C IMS Uncertainty

This appendix describes an experiment conducted in the shop facility at NIST on July 7, 2011 using a robot to determine the uncertainties of an Independent Measurement System (IMS). The IMS uses a network of infrared cameras and special retro-reflective targets to triangulate position. The goal of this experiment was to verify that the IMS solution was sufficiently accurate to verify the positions provided by human tracking software integrating data from the laser measurement sensor and the behavior of the speed and separation monitoring (SSM) software to be used later on the robot.

NIST evaluated the robot accuracy according to the ISO 9283:1998(E) "Manipulating industrial robots—Performance criteria and related test methods." The largest errors were less than 11 mm that occurred at the highest speeds tested (2 m/s) and the highest load tested (20 kg). The reported accuracy of the robot for the IMS tests being described now where the robot moves at 0.5 m/s carrying a load that consists primarily of a hollow plastic tube obviously less than 20 kg is around 2 mm.

Both the robot and the IMS provide relative time stamps from independent and unknown epochs, i.e., seconds since that particular device was powered. In order to compare points we first need to convert the time stamps to some common epoch. To do this an offset is added to the device time stamps. The offset is the mean difference between the device time stamp and the Network Time Protocol (NTP) synchronized time on the computer receiving the data measured as seconds since 00:00:00 UTC, January 1, 1970.

In addition to applying timestamps based on a common epoch, it is also necessary to adjust the data for the different sampling frequency as well as the different times the logging was started and ended.

From the robot 2055 points were recorded over a 38.3 second period (sample rate = 53 Hz or 18.9 ms sample period) and 5392 points were recorded from the IMS over a 53.9 second period (sample rate = 100 Hz or 10 ms sample period). The reason they were recorded for different periods is that the data was recorded on separate computers that had to have the logging programs manually started and stopped. Linear interpolation is used to achieve a one-to-one correspondence between points with different time stamps.

The IMS coordinate system is established not by the location of any of the cameras but based on the position of the calibration square target when a picture of it is taken. The calibration square target was placed on the floor approximately under the center of the robot's work volume and in the FOV of all three cameras. The exact transform between where the calibration square was placed and the base of the robot is unknown. Also the transform between the TCP reported by the robot and the position of the target on the tube being held by the robot is also unknown. In order to compare points reported by the robot with positions of the target measured by the IMS, it is necessary for both sets to be transformed into a common coordinate system. An automated hand-eye technique [10] and a manual overlay technique were used to estimate the transforms.

The robot was programmed to follow a path that kept it at a fixed height and moved around a box with a zigzag on one side covering the area expected to be used for later (SSM/human tracking) experiments. Since the SSM/human tracking experiments will primarily be concerned with the 2D X and Y locations, the program did not include changes in Z or the roll/pitch/yaw rotations. To validate the IMS for more general 3D applications, the program should be modified to include a broader sample of z positions and rotations.

Figure 16 shows the original untransformed IMS data in red as well as the robot data in green and the IMS data transformed into robot coordinates with two different methods in cyan (hand-eye [10]) and purple (manual overlay).

23

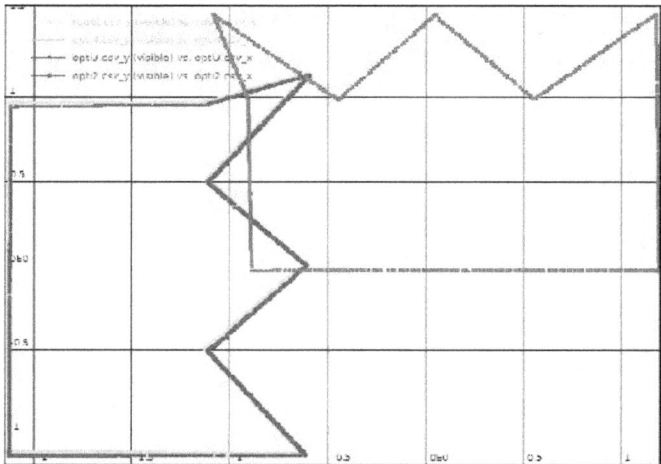

Figure 16. Overhead XY Plots: 0.5m grid, green = robot, red = IMS original, purple = IMS transform (hand-eye), cyan = IMS transform (manual overlay)

Appendix D <u>Time to contact</u>

Time to contact is a useful calculation for determining the potential for contact between the robot and the human. Figure 17 illustrates graphically a contact situation between a robot and a person. The vectors $\mathbf{p_r}$, $\mathbf{v_r}$, $\mathbf{p_h}$, and $\mathbf{v_h}$ are the position and velocity vectors of the robot and the human.

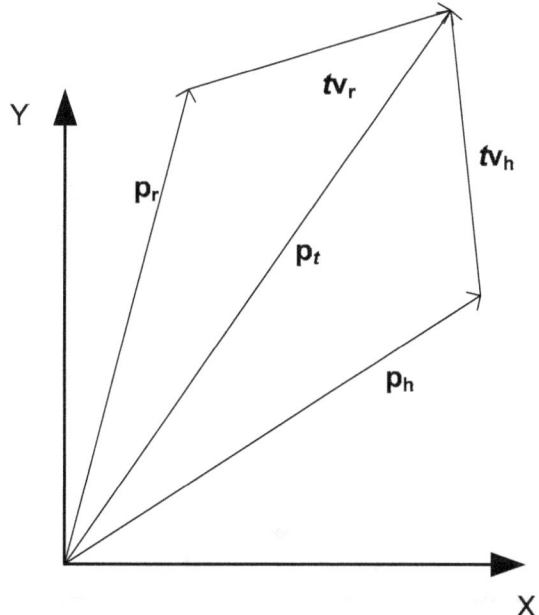

Figure 17 Robot and human position and velocity

Using $\mathbf{v} = \mathbf{v_h} - \mathbf{v_r}$ and $\mathbf{p} = \mathbf{p_h} - \mathbf{p_r}$, the closing speed is defined as which is the projection (i.e., dot product) of the closing velocity vector on to the normalized vector ($\hat{p}$) defining the direction from the robot to the human. Using v_c one can determine the following values:

Time to contact:
$$t = \frac{\|p\|}{v_c}$$

Distance to contact:
$$t \times v_c$$

Robot position at contact ($\mathbf{p}_t$):
$$p_r + t v_r$$